Rajni Garg, R.D. Singh

Effect of addition of crown ether on the micellar behavior of Dodecyltrimethylammonium Chloride in Aqueous media

GRIN Verlag

Bibliografische Information der Deutschen Nationalbibliothek:

Die Deutsche Bibliothek verzeichnet diese Publikation in der Deutschen National-
bibliografie; detaillierte bibliografische Daten sind im Internet über http://dnb.d-
nb.de/ abrufbar.

Imprint:

Copyright © 2011 GRIN Verlag GmbH
Druck und Bindung: Books on Demand GmbH, Norderstedt Germany
ISBN: 978-3-656-08943-8

This book at GRIN:

http://www.grin.com/en/e-book/183933/effect-of-addition-of-crown-ether-on-the-
micellar-behavior-of-dodecyltrimethylammonium

GRIN - Your knowledge has value

Der GRIN Verlag publiziert seit 1998 wissenschaftliche Arbeiten von Studenten, Hochschullehrern und anderen Akademikern als eBook und gedrucktes Buch. Die Verlagswebsite www.grin.com ist die ideale Plattform zur Veröffentlichung von Hausarbeiten, Abschlussarbeiten, wissenschaftlichen Aufsätzen, Dissertationen und Fachbüchern.

Visit us on the internet:

http://www.grin.com/

http://www.facebook.com/grincom

http://www.twitter.com/grin_com

Effect of addition of crown ether on the micellar behavior of Dodecyltrimethylammonium Chloride in Aqueous media

[*] Rajni Bala[a] and R.D.Singh[a]

a Department of Chemistry, Gurukula Kangri University, Haridwar (India)

Abstract

The micellar properties of cationic dodecyltrimethylammonium chloride (DTAC) in aqueous media in the presence of 15-crown-5ether (CR) have been investigated by conductivity measurements over the temperature range 288.15-308.15 K. The results of the ternary DTAC/CR/W system were analysed in comparison with the reported results of binary DTAC/W system. The critical aggregation concentration (cac) and degree of ionization (β) of the micelles were determined from the conductivity measurements at different temperatures. Thermodynamic parameters (ΔG_m^0, ΔH_m^0 and ΔS_m^0) for the micellar system were estimated by applying the charged pseudo-phase separation model. Micellisation was found to be spontaneous and entropy-driven.

Keywords: Crown ether; Dodecyltrimethylammonium chloride; Critical aggregation concentration; Degree of ionization

Introduction

Surfactant organized assemblies have great potential applications in day to day life [1]. Surfactants have micellar properties, which are effected by addition of small amount of electrolytes, non-polar and polar organic compounds. The critical aggregation concentration (cac) results from the hydrophobic interactions between the non- polar parts, which forms the core of the micelles and a repulsion interaction between the polar head group. During the last decades, the study of the behavior of ionic surfactants has received much attention. The effect of different kinds of additives including macrocycles, on the micellisation has also been widely studied. In some applications, macrocyclic compounds are used along with the surfactants due to counterion complexation. Crown ethers are a versatile class of macrocyclic ligands. The property of complex formation along with the selectivity shown by crown ethers towards cations distinguishes them from most non-cyclic ligands. The formation of complexes between the crown ether cavity and the counterions is expected to lead to significant alterations of the micellar properties. However, not much investigation is reported in this field [2-16].

In the present work, efforts have been made to study the effect of addition of crown ether (15-crown-5) with wider temperature variations on the *cac* of aqueous solutions of cationic dodecyltrimethylammonium chloride (DTAC) using conductivity technique. In addition, the thermodynamic parameters, ΔG_m^0, ΔH_m^0 and ΔS_m^0 have also been estimated and analyzed in aqueous media.

Experimental

Materials

The Crown ether (15-crown-5) from Fluka (purity greater than 99%) was used as received. Dodecyltrimethylammonium chloride (DTAC) (purity greater than 99%) was purchased from Sigma and was recrystallised from ethanol- ethyl acetate mixtures. It was dried in a vacuum oven at 60°C for two days. Water used for the preparation of samples was deionized and triply distilled (conductivity lower than $3\,\mu$ s).

Methods

The conductivity of ternary mixtures was measured in a thermostatic glass cell with two platinum electrodes and Pico conductivity Meter from Lab India. The conductivity meter was calibrated by measuring the conductivity of the solutions of potassium chloride (Merck, purity > 99%) of different concentrations (0.001 M, 0.01M and 0.1M). Electrodes were inserted in a double walled glass cell containing the solution. The glass cell was connected to the thermostat controlled to better than ±0.01K temperature variation, read on Beckman thermometer set at the working temperature. The cell constant of the cell used was $1cm^{-1}$. The measurement of conductivity was carried out with an absolute accuracy up to ±3%. The solutions were prepared by weight using an analytical balance with an accuracy of $±1.10^{-4}$ g. The conductivity measurements were made at different temperatures viz 288.15-308.15 K for ternary DTAC/CR/W systems, as a function of surfactant concentration with fixed concentration of crown ether (0.01M).

Results and Discussion

Conductivity technique has been found to be a useful tool for studying the association behavior of various systems [17-25]. Ternary DTAC/CR/W systems have been characterized through conductivity measurements at various temperatures (288.15–308.15 K) in order to evaluate the critical aggregation concentration, cac and the degree of counterion dissociation, β. The intersection of the two straight lines of the conductivity-concentration plots above and below the change in the slope gives the cac while the ratio between the slopes of the postmicellar region to that in the premicellar region gives the degree of counterion dissociation, β [20-25].Figure 1 shows conductivity, κ, as a function of DTAC concentration for DTAC/CR/W solutions at a constant concentration of CR at different temperatures.

The conductivity plots versus concentration show a gradual increase in conductivity with increase in temperature. The temperature dependence values of cac for ternary solutions along with the reported values for binary systems are also shown in table 1. The values for the ternary systems when compared with that of binary systems indicate the ease in micellisation. Figure 2 shows the variance in cac with increase of temperature. There is a gradual decrease in cac in the temperature range investigated. This observed decrease may be probably due to decrease in the degree of hydration of the hydrophilic group with increase in the thermal energy of the molecular entities which favors the micellisation. The values of degree of counterion dissociation, β are tabulated in table 1.

The regular increase in β with increase in temperature may be attributed to the decrease in the charge density at the micellar surface caused by the decrease in the aggregation number of the micelle. Such behavior indicates that micelles of smaller aggregation number and/or higher degree of ionization, β, are formed at higher temperatures as expected [20-25]. Figure 3 shows the variation of β as a function of temperature. The linear dependence of β with temperature variation in this class of surfactant is in accordance with the findings [23].

In accordance with the charged pseudo-phase separation model, the standard free energy of micellisation per mole of surfactant, ΔG_m^0, was calculated from the relation,

$$\Delta G_m^0 = (2 - \beta)RT \ln X_{cac} \tag{1}$$

where R is the gas constant, T is the temperature, and X_{cac} is the cac value expressed in terms of mole fraction [23].

Figure 4 shows the relationship between the standard free energy of micelle formation and temperature. The values of ΔG_m^0 at various temperatures are listed in table 2. For amphoteric and ionic surfactants, ΔG_m^0 has been reported to be between - 23 and -42 kJmol^{-1} at 298.15 K [25]. The free energy values for DTAC/ CR/ W fall within this range. ΔG_m^0 values decrease with increase in temperature indicating the spontaneous micellisation and decrease in hydrophobic interaction.

The standard enthalpy of micellisation (ΔH_m^0) has been derived from the van't Hoff equation

$$\Delta H_m^0 = -RT^2 (2 - \beta) \left(\frac{d\ln X_{cac}}{dT} \right) \tag{2}$$

For the ternary systems, the term $\left(\frac{d\ln X_{cac}}{dT} \right)$ was calculated by fitting the ln X_{cac} versus temperature data to a second order polynomial and differentiation. The standard enthalpy of micellisation (ΔH_m^0) is positive as listed in table 2. Figure 5 shows variation in enthalpy change with temperature for both the systems. ΔH_m^0 is found to increase with temperature for the studied systems indicating the endothermic process of micellisation.

The entropy of micellisation, ΔS_m^0 was determined from the equation

$$\Delta S_m^o = \left(\frac{\Delta H_m^o - \Delta G_m^o}{T} \right)$$ (3)

The estimated values of ΔS_m^0 are listed in table 2 and are plotted in figure 6. The ΔS_m^0 values for the ternary systems are positive at the range of temperature studied. This indicates that the micellisation process is entropy dominated as expected. The ΔH_m^0 values are much smaller than the values of $T \Delta S_m^0$. Therefore the micellisation process is governed primarily by the entropy gain associated with it. This may be due to destruction of structured water molecules around a hydrophobic chain with the interaction of crown ether molecules with the surfactant micelles.

Conclusions

Effect of addition of crown ether on the micellar behavior of cationic DTAC in aqueous media has been investigated with the help of conductivity measurements over the temperature range of 288.15-308.15K. A linear fall in *cac* was observed in the ternary systems probably due to an increase in the hydrophobic character of molecule (DTAC) along with the interaction with crown ether. Thermodynamics of the system reveals that the micellisation was found to be entropy- driven. However more investigation is required to explore such novel systems.

6

References

1. M. J.Rosen, Surfactant and Interfacial Phenomena, Wiley-Interscience pub. 1989, Chapter 1.

2. G. Gokel: Crown ethers and Cryptands, The Royal Society of Chemistry, Cambridge (1991).

3. P. Baglioni, L. Kevan, J. Chem. Soc.., Faraday Trans.1. 84 (1988) 467.

4. P. Baglioni, L. Kevan, E. Rivera-Minten, J. Phys .Chem. 92 (1988) 4726.

5. H. J. D. McManus, Y. S. Kang, L. Kevan, J. Phys .Chem. 97 (1993) 255.

6. M.S. Bakshi, R. Crisantino, R. De Lisi, S. Milioto, Langmuir, 10 (1994) 423.

7. P .Stilbs, J. Colloid Interface Sci. 87 (1982) 385.

8. P .Stilbs, J. Colloid Interface Sci. 94 (1983) 463.

9. D. F. Evans, R. Sen, G.G. Warr, J. Phys .Chem. 90 (1986) 5500.

10. D. F. Evans, R. Sen, G.G. Warr, J.B. Evans, J. Phys .Chem. 92 (1988) 784.

11. G.capuzzi, E. Fratini, F. Pini, L. Dei, P. Lo Nostro, A. Casnati, R. Gilles, P. Baglioni, Colloids Surf. A 167 (2000) 105.

12. G.capuzzi, E. Fratini, F. Pini, P. Baglioni, A. Casnati, Teixeira, Langmuir, 16 (2000) 105.

13. G. Ilgenfritz, R. Schneider, E. Grell, E. Lewitzki, H. Ruf, Langmuir 20 (2004) 1620.

14. E. Caponetti, D. C. Martino, L. Pedone, J. Appl. Cryst. 36 (2003) 753.

15. E. Caponetti, D. C. Martino, L. Pedone, Langmuir, 20 (2004) 3854.

16. E. Caponetti, D. C. Martino, M. A. Floriano, R. Triolo, G. D. Wignall, Langmuir 11 (1995) 2464.

17. S.P. Moulik, Md.E. Haque, P.K. Jana, A.R. Das, J. Phys. Chem. 100 (1996) 701.

18. S.K. Mehta, K.K. Bhasin, Renu Chauhan, Shilpee Dham, Colloids Surf. 255 (2005) 153.

19. A. Gonzalez-Perez, J. Czapkiewicz, J.L. Del Castillo, J.R. Rodriguez, Colloid Polym Sci. 281 (2003) 556.

20. A. Gonzalez-Perez, J. Czapkiewicz, J.L. Del Castillo, J.R. Rodriguez, Colloids Surf. 193 (2001) 129.

21. A. Gonzalez-Perez, J.L. Del Castillo, J. Czapkiewicz, J.R. Rodriguez, Colloid Polym Sci. 280 (2002) 503.

22. J.R. Rodriguez, A. Gonzalez-Perez, J.L. Del Castillo, J. Czapkiewicz, J. Colloid Interface Sci. 250 (2002) 438.

23. A. Gonzalez-Perez, J.L. Del Castillo, J. Czapkiewicz, J.R. Rodriguez, Colloids Surf. 232 (2004) 183.

24. A. Gonzalez-Perez, J. Czapkiewicz, J.L. Del Castillo, J.R. Rodriguez, J. Phys. Chem. B 105 (2001) 1720.

25. K.H. Kang, H.U. Kim, K.H. Lim, Colloids Surf. 189 (2001) 113.

Table 1. Values of *cac* and degree of ionisation (β) for DTAC/ CR/ W and DTAC/W at various temperatures.

| T | DTAC/ CR/ W | | DTAC/W[a] | |
| | cac | β | cac | β |
(K)	(mol kg^{-1})		(mol kg^{-1})	
288.15	0.0226	0.360	0.0231	0.328
293.15	0.0214	0.382	0.0221	0.354
298.15	0.0201	0.412	0.0213	0.389
303.15	0.0194	0.445	0.0204	0.421
308.15	0.0185	0.485	0.0196	0.450

[a]Ref. 18

Table 2. Values of the free energy of micellisation, ΔG_m^0, enthalpy of micellisation, ΔH_m^0 and entropy of micellisation, ΔS_m^0 for DTAC/ CR/ W at various temperatures.

| T | ΔG_m^0 | ΔH_m^0 | ΔS_m^0 |
(K)	(kJ mol^{-1})	(kJ mol^{-1})	(kJ mol^{-1} K^{-1})
288.15	-29.56	18.91	0.168
293.15	-32.16	19.31	0.176
298.15	-34.20	19.61	0.180
303.15	-35.82	19.85	0.184
308.15	-37.03	19.98	0.185

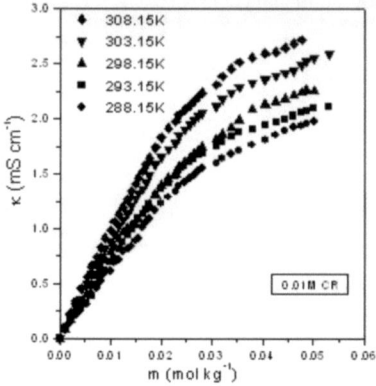

Fig.1. Temperature dependence of specific conductivity κ versus molality m

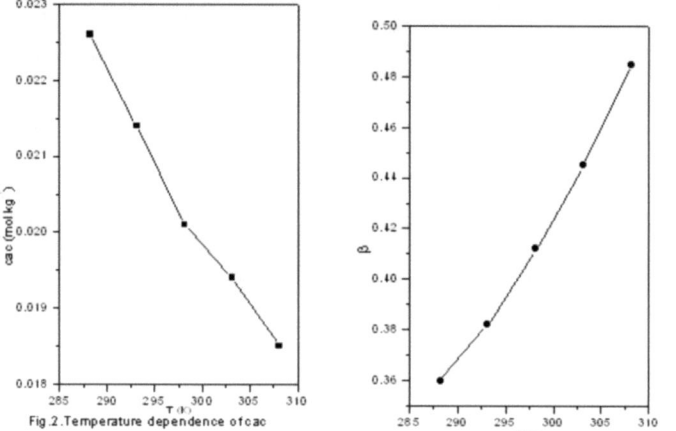

Fig.2.Temperature dependence of cac

Fig.3 Temperature dependence of degree of ionization (β)

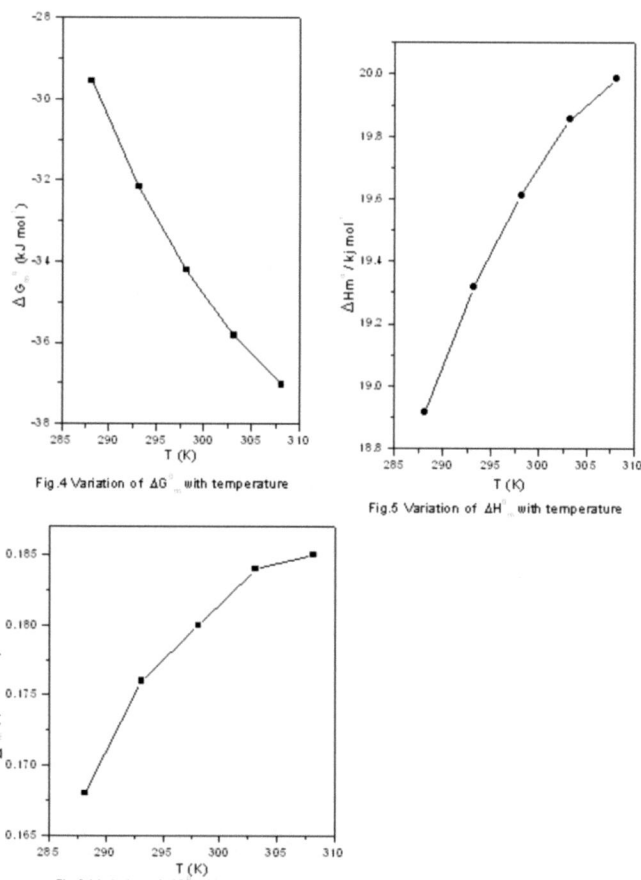

Fig.4 Variation of $\Delta G^{\ominus}_{m}$ with temperature

Fig.5 Variation of $\Delta H^{\ominus}_{m}$ with temperature

Fig.6 Variation of $\Delta S^{\ominus}_{m}$ with temperature